discovermore

Forms of Energy

What Is Sound Energy?

Virginia Elliott

Britannica Educational Publishing

IN ASSOCIATION WITH

THE ROSEN PUBLISHING GROUP

Published in 2026 by Britannica Educational Publishing (a trademark of Encyclopædia Britannica, Inc.) in association with The Rosen Publishing Group, Inc.
2544 Clinton Street, Buffalo, NY 14224

Distributed exclusively by Rosen Publishing.
To see additional Britannica Educational Publishing titles, go to rosenpublishing.com.

Editor: Danielle Haynes
Book Design: Michael Flynn

Photo Credits: Cover PV productions/Shutterstock.com; (series background) Dai Yim/Shutterstock.com; p. 4 Ground Picture/Shutterstock.com; p. 5 shodography/Shutterstock.com; p. 7 (top) Andrew Berezovsky/Shutterstock.com; p. 7 (bottom) sanjayart/Shutterstock.com; p. 9 Master_shifu/Shutterstock.com; p. 10 buradaki/Shutterstock.com; p. 11 VectorMine/Shutterstock.com; p. 12 Ververidis Vasilis/Shutterstock.com; p. 13 Aree_S/Shutterstock.com; p. 15 (top) Lipik Stock Media/Shutterstock.com; p. 15 (bottom) Photo Oz/Shutterstock.com; p. 16 metamorworks/Shutterstock.com; p. 17 eric1207cvb/Shutterstock.com; p. 18 DC Studio/Shutterstock.com; p. 19 Thomas Nuehnen/Shutterstock.com; p. 21 (top) SpeedKingz/Shutterstock.com; p. 21 (bottom) TarikVision/Shutterstock.com; p. 22 Pepermpron/Shutterstock.com; p. 23 Paolo Trovo/Shutterstock.com; p. 24 Monkey Business Images/Shutterstock.com; p. 25 edwardolive/Shutterstock.com; p. 26 https://commons.wikimedia.org/wiki/File:Alexander_Graham_Telephone_in_Newyork.jpg; p. 27 Thales Santos/Shutterstock.com; p. 28 paffy/Shutterstock.com; p. 29 riopatuca/Shutterstock.com.

Cataloging-in-Publication Data

Names: Elliott, Virginia (Children's writer).
Title: Sound energy / Virginia Elliott.
Description: New York : Britannica Educational Publishing, in association with Rosen Educational Services, 2026. | Series: Discover more: forms of energy | Includes glossary and index.
Identifiers: ISBN 9781641905756 (library bound) | ISBN 9781641905749 (pbk) | ISBN 9781641905763 (ebook)
Subjects: LCSH: Sound--Juvenile literature. | Sound-waves--Juvenile literature.
Classification: LCC QC225.5 E45 2026 | DDC 534--dc23

Manufactured in China

Some of the images in this book illustrate individuals who are models. The depictions do not imply actual situations or events.

CPSIA Compliance Information: Batch #QWBRIT26. For further information contact Rosen Publishing at 1-800-237-9932.

Contents

Energy Basics

Energy is the ability to do work. Have you ever heard someone describe a person by saying they have a lot of energy? Usually, when someone says that they are talking about a person who moves around a lot. It's a good description because things and people need energy to move.

Energy is another word for power. It makes things move or work and makes living things grow. Energy can be divided into potential energy and kinetic energy. Potential energy is power that is stored, to use later. Kinetic energy is power in action, or moving energy.

You need energy to be able to run, walk, and even breathe.

If you've ever seen a type of drum called timpani, you'll notice its surface vibrates when it's hit.

There are many different kinds of energy, such as electricity, heat, and light. Sound energy is made when an object, or part of an object, **vibrates**. The vibrations move the air around it to make sound waves. These are what we hear, but we don't always see them.

WORD WISE

SOMETHING THAT VIBRATES HAS A SHAKING MOVEMENT.

Listen Up

Even though you usually never see it, all sound travels in the same way. It starts at a source, such as a bell that is rung or an object that is hit. A medium, such as air, carries the sound waves. Finally, a receiver detects the sound.

Your receiver is your ear. The human ear has three parts. Each part has its own job. The outer ear collects sound waves and contains the eardrum. The eardrum vibrates when the waves hit it. The middle ear sends the vibrations to the inner ear over three small bones. Finally, the vibrations go through a fluid in the inner ear, making waves that send a sound signal to your brain.

Sound waves aren't like waves in the ocean, though. They bend in and out quickly, to push and pull the air around them. Sound waves are called longitudinal waves, while waves in the ocean, called transverse waves, go up and down.

Sound waves can also carry through water, as can be seen in this speaker causing water to jump.

compareandcontrast

How are the outer ear, middle ear, and inner ear alike? How are they different?

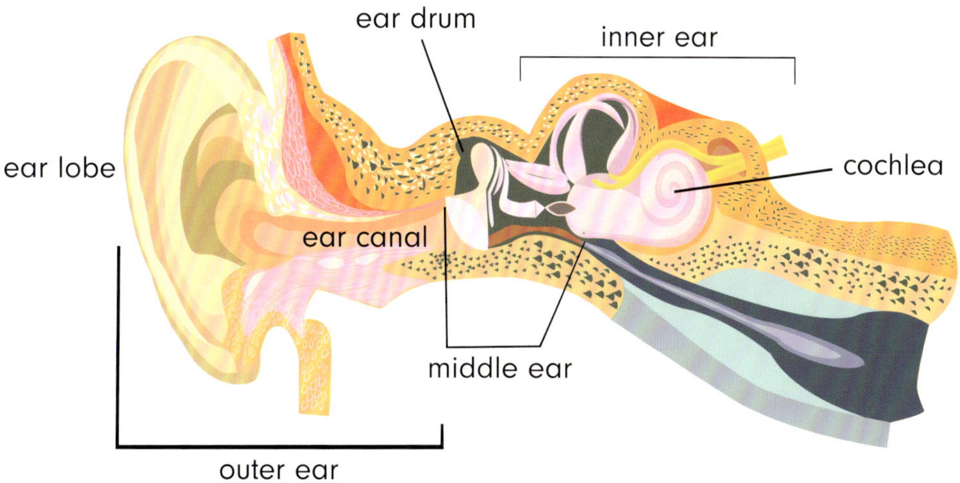

The fluid in the inner ear is in a spiral-shaped part of the ear called the cochlea.

Properties of Sound

So, how do you describe sound? All sounds are soft or loud, high or low, and pleasant or unpleasant. This is because every sound has several different properties. Scientists use systems of measurement to describe each property of sound.

"Intensity" is a word that is used to describe a sound's volume, or how soft or loud the sound is. Intensity is the strength of the vibrations producing a sound. When a door is shut carefully, the vibrations made by that movement are weak and don't push the air with much force. This produces a quiet sound. When a door is slammed, strong vibrations are made. This creates a loud sound. Intensity is measured in decibels (dB). A measurement of 0 dB is the softest sound you can hear, while thunder measures at 120 dB. Anything higher than 120 dB can be dangerous for your ears.

Consider This

Would a police siren measure at a higher or lower number of decibels than a concert?

Noise Level Decibel Chart

Sound	Decibels	Level
fireworks, gunshot	140 dB	painful, dangerous noise levels
jet engine	130 dB	
rock concert, siren	120 dB	
leaf blower	110 dB	Very loud, dangerous over 30 minutes
motorcycle	100 dB	
hairdryer	90 dB	
lawn mower	80 dB	Loud
office noise	70 dB	
normal conversation	60 dB	Moderate
moderate rainfall	50 dB	
average room noise	40 dB	Soft, safe noise levels
whispering	30 dB	
rustling leaves	20 dB	
breathing	10 dB	
hearing threshold	0 dB	

What can you hear right now? About how many decibels do you think it is?

Another way we measure sound is pitch. Pitch is how high or low a sound is. Faster vibration causes a higher pitch, while slower vibration causes a lower pitch. Wavelength, the distance from the peak of one wave to the peak of the next, is one way to measure vibration. Sounds with shorter wavelengths have a higher pitch and **frequency**. The unit of measurement used to record frequency is the hertz (Hz). One hertz is a complete vibration back and forth.

How loud a sound wave is doesn't affect the speed at which it travels. At room temperature, 70°F (21°C), sound travels 1,129 feet (334 meters) per second. That means that it takes sound about 5 seconds to travel 1 mile (1.6 kilometers). The warmer the air, the faster sound will travel. Sound waves travel faster through water than they do through the air. They travel fastest through solids such as metal and stone.

Sound waves need to be able to move particles to travel. This is why sound doesn't travel in the vacuum of outer space.

Amplitude

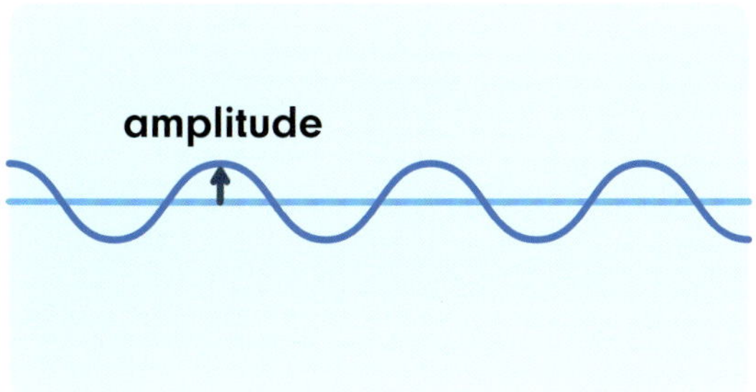

low amplitude - quiet sound

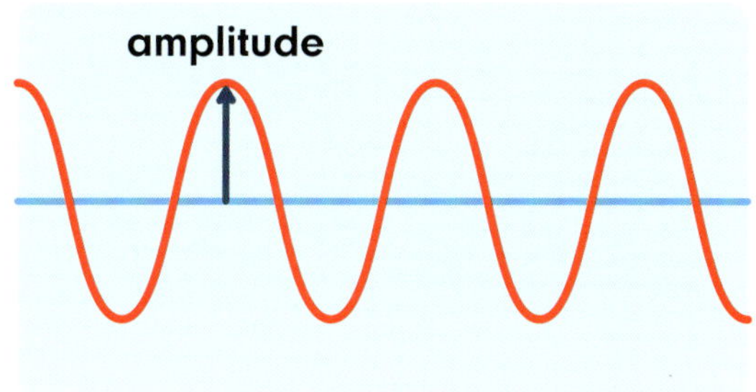

high amplitude - loud sound

Pitch

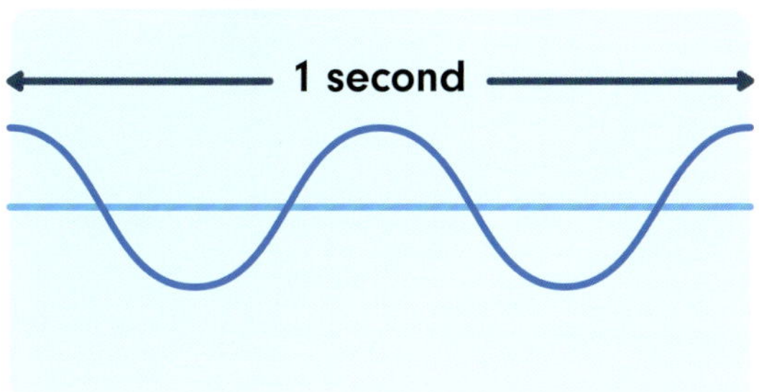

low frequency - low pitch - low sound

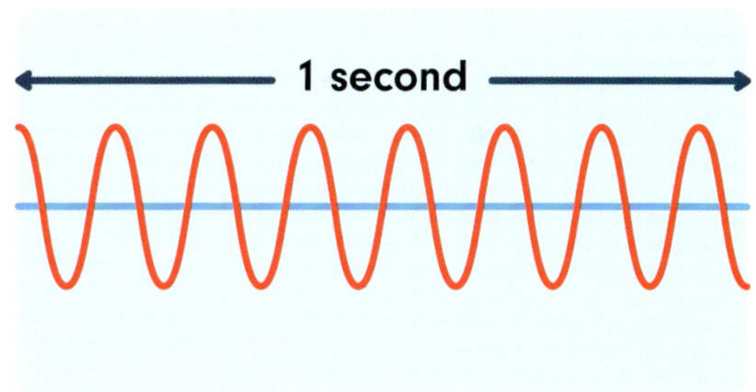

high frequency - high pitch - high sound

The louder, or greater amplitude, a sound has, the stronger the wavelengths are. High-pitch sounds create more frequent waves than low-pitch sounds.

WORD WISE

FREQUENCY IS THE NUMBER OF COMPLETE VIBRATIONS PER SECOND.

Reflecting Sound

Sound waves move around a lot more than you might think. They can echo, or bounce off of, other surfaces. This changes the way the sound is received.

Auditoriums are built to focus or strengthen sound. The shape and size of the room affects how sound reflects, or bounces around. Auditoriums usually have high, curved ceilings, and walls that move out at an angle from the stage. Some have panels hanging from the ceiling to provide more surfaces for sound waves to bounce off.

As technology has advanced, newer auditoriums have made advances to improve how sound travels.

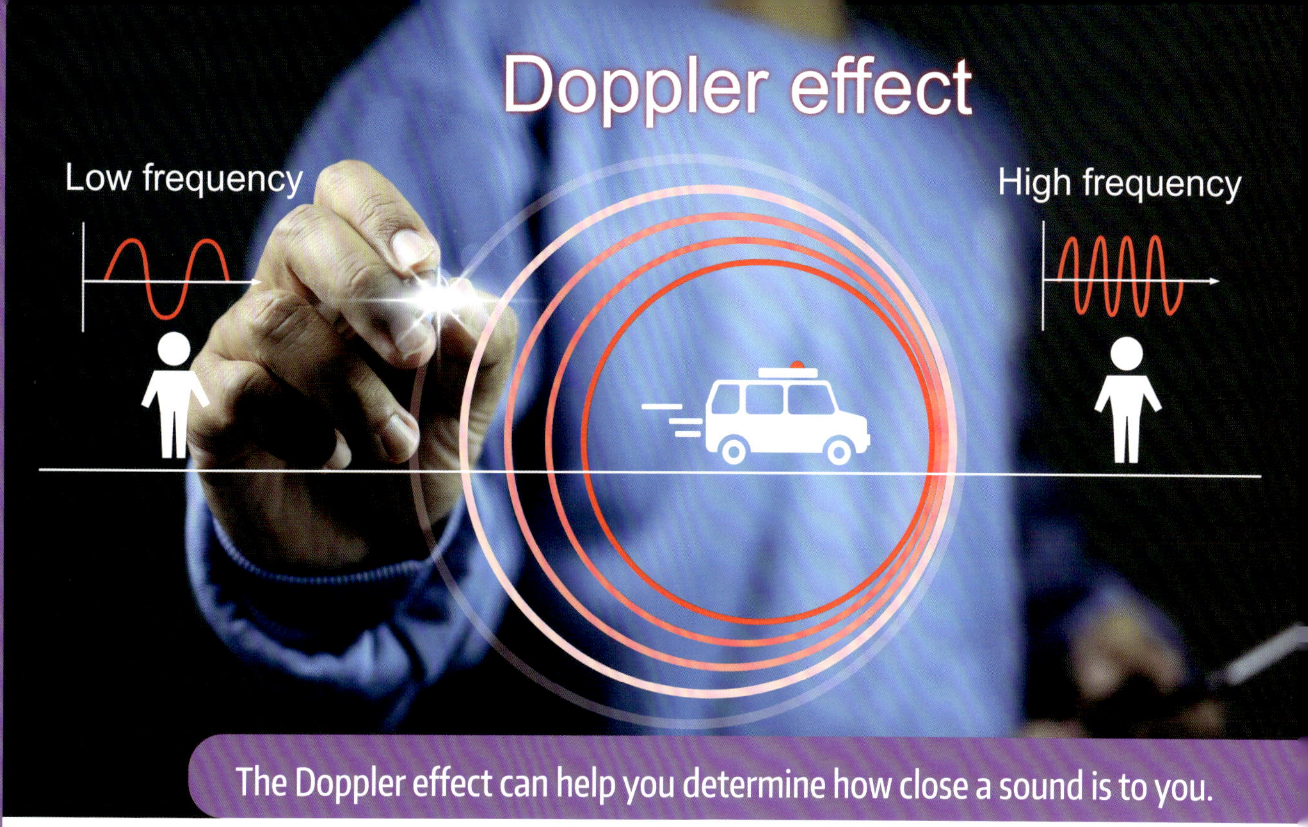

The Doppler effect can help you determine how close a sound is to you.

When the source of a sound is moving, the sound changes. As the source of the sound, like an ambulance siren, gets closer to you, it sounds like it has a higher pitch. This is because the wavelengths are shortened. As it moves away, the wavelengths get longer, and so the pitch is lower. This is called the Doppler effect. Scientists also use the Doppler effect to predict the weather.

WORD WISE

AUDITORIUMS ARE PLACES WHERE PEOPLE SEE PLAYS AND LISTEN TO CONCERTS.

Music to Your Ears

Musicians can produce sound through vibrations of their instruments. Different types of instruments produce vibrations in different ways.

String instruments, such as violins, make sounds when a string is pulled, hit, or rubbed with a bow. The sound waves from the vibrating strings bounce around the inside of the body of the instrument. The pitch of the sound depends on how long the string is and the strength of the pull upon the string.

Flutes and other wind instruments create sound when air is blown into them. A flute player blows across an opening in a tube. The air that goes in the tube bounces around, producing sound.

Percussion instruments, such as drums, are hit to produce sound. The tightness of the drum skin, the size and shape of the drum, and how it is hit change what types of sounds are made. All of these vibrations then travel to your ears, making music.

The vibrations from the bow moving across the strings of a violin create different musical notes.

compare and contrast

How are the sounds of these types of instruments the same? How are they different?

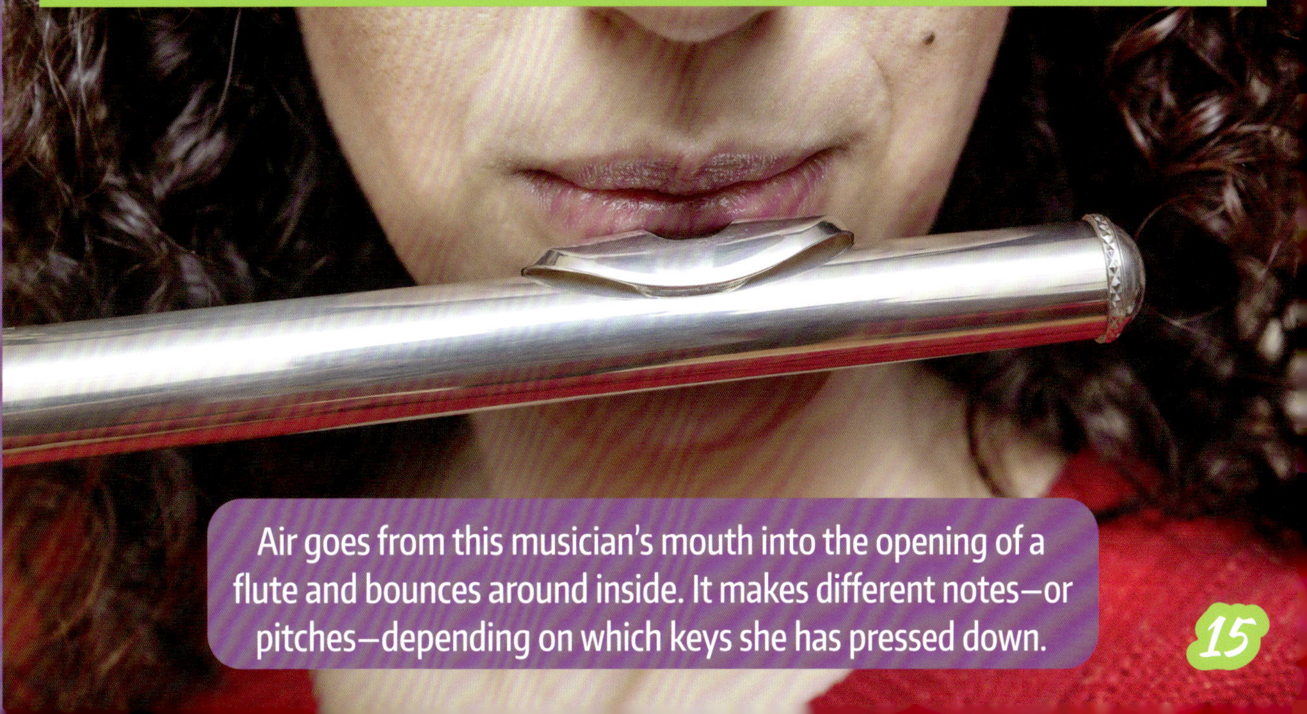

Air goes from this musician's mouth into the opening of a flute and bounces around inside. It makes different notes—or pitches—depending on which keys she has pressed down.

Make Some Noise

Sound gets weaker the farther it travels. When we want sounds to reach a lot of people, we use something to amplify the sound, or make it louder. Two devices used to amplify sound are microphones and speakers.

A microphone is a receiver, like your ear. It picks up sound waves. When you speak or sing into a microphone, the sound waves hit a piece of material that captures vibrations. It then turns that sound energy into electrical energy, which can be delivered to a speaker. Microphones are found in telephones, computers, and hearing aids.

This man uses a microphone during a presentation so many people in a room can hear him from far away.

Large speakers hang on either side of this stage to amplify the performers' music to a large area.

A speaker works the opposite way. The electrical energy caught by the microphone is changed back into vibrations through magnets. The magnets are attached to a cone-shaped piece. A good speaker will have different sized cones to reflect different frequencies of sound. Speakers can be large, such as those found at concerts, while those in stereos, headphones, and televisions are much smaller.

Consider This

In what situations might people need a way to amplify sound?

Capturing and Listening to Sound

Microphones and speakers are also needed in sound, or **audio**, recording. This is the storage of sounds so that they can be heard again. In professional and home recording studios, people use special equipment to record music so that they can listen to it whenever they want.

Recording devices can preserve sound in a number of formats. The earliest formats were phonograph records and magnetic tape. A phonograph record has a groove with patterns cut into it that represents the waves of sound. Magnetic tape represents sound with wavelike patterns of magnetized particles.

Sound engineers use lots of high-tech equipment to record and manipulate sound for musicians.

The needle of a phonograph moves through tiny grooves on a large disc, or record. A speaker amplifies the music.

Most sound-recording devices today are digital devices. They store sound as a long series of numbers that describe the sound waves. CDs store this information as a pattern of tiny pits, or holes, that are created and read by a laser beam. These days digital information is most often stored in personal computers, phones, or portable devices called digital audio players. Some music lovers still enjoy listening to music on vinyl records, though.

WORD WISE

AUDIO IS THE SENDING, RECEIVING, OR REPRODUCING OF SOUND.

Animals and Sound

Human hearing is very different from animal hearing. Some sounds are too high or low for us to hear because of how our ears are shaped. People can typically hear sounds between 20 and 20,000 Hz. A turtle can hear only between 20 and 1,000 Hz because it doesn't have outer ears. A dog or cat, on the other hand, can hear more than double the range of frequencies we can. This is why they can tell when someone is coming to your door before they ring the bell.

Some animals use a process called echolocation to help them get around. A bat makes a high-pitched noise while it flies, which people can't hear. The sound echoes back to the bat and helps it figure out how close it is to other objects. Bats use this information to help them find food. Though bats rely on this technique to help them navigate, they do have functional eyes and can see!

Some people use special whistles that only dogs can hear to train their pets.

compareandcontrast

How are a dog's ears like a person's? How are they different?

Dolphins, like bats, use echolocation to find food. They can tell how far away an object is by the amplitude of the returning sound.

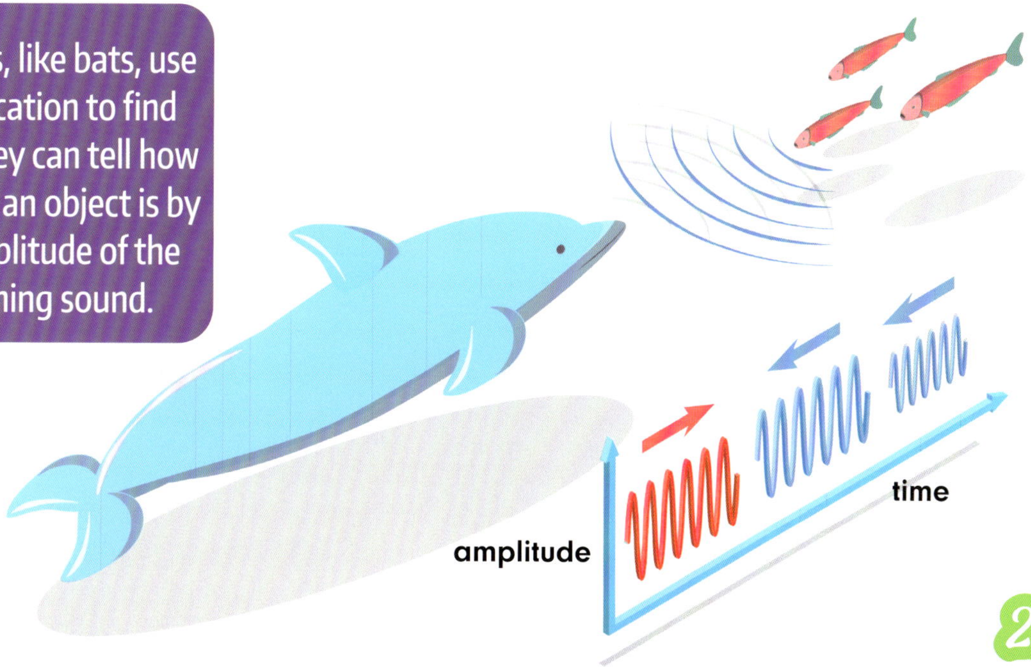

Seeing with Sound

Dolphins and bats aren't the only ones to use echolocation. Humans have created a technology that mimics this ability. Sonar systems use sound waves to detect objects and figure out their size, their shape, and even how they are moving. Some systems send out the sound waves and pick up the reflections. Others receive sound waves sent by moving objects.

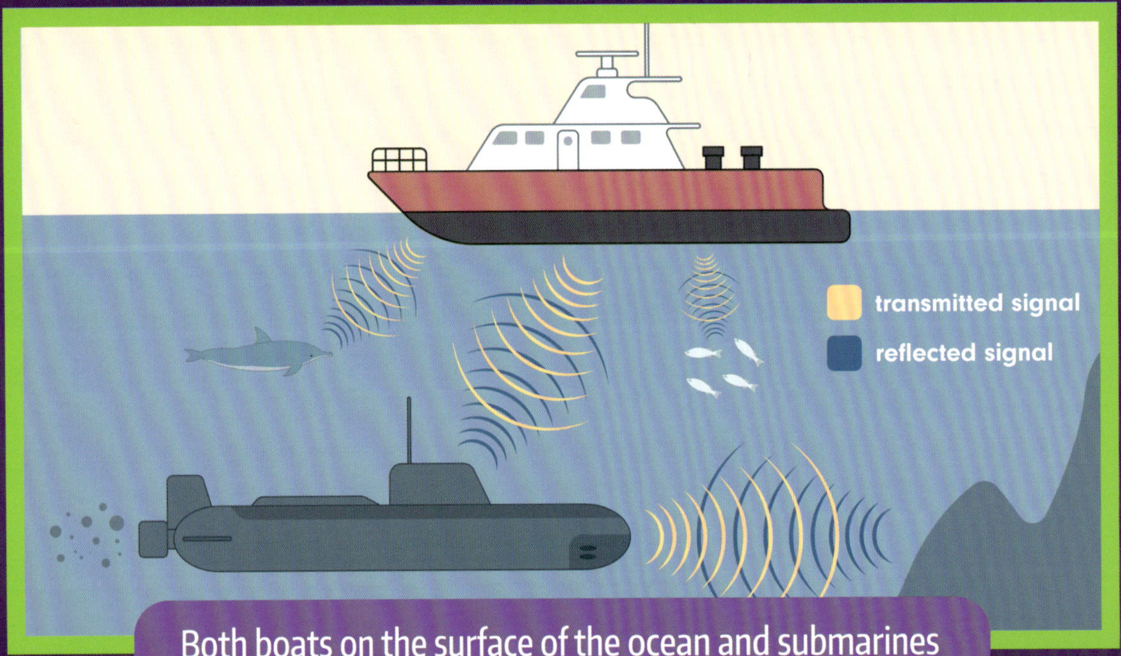

Both boats on the surface of the ocean and submarines down deep use sonar to "see" what's around them.

Investigators use this sonar equipment on board a boat to try to find large animals underwater in Scotland's Loch Ness.

The word "sonar" stands for "sound navigation and ranging." Sonar technology is useful in situations where it is hard to see, such as underwater. In the ocean, sound waves can travel farther than light waves. Sonar can be used to study sea animals, find shipwrecks, and make maps of the ocean floor. Fishermen can use it to find large groups of fish to catch. Some people have even used the technology to try to prove whether the Loch Ness Monster is real!

Consider This

What could be some other uses for sonar technology?

Sound Waves in Medicine

The sounds we cannot hear with our ears are useful to us in many ways. In medicine, doctors use a technology to produce sounds above 20,000 Hz called ultrasound. A special wand is used to send sound waves through the skin and into the body. The waves that reflect back are turned into a picture, so that the doctor can see what is inside the body. Ultrasound is used to detect illnesses or examine babies before they are born.

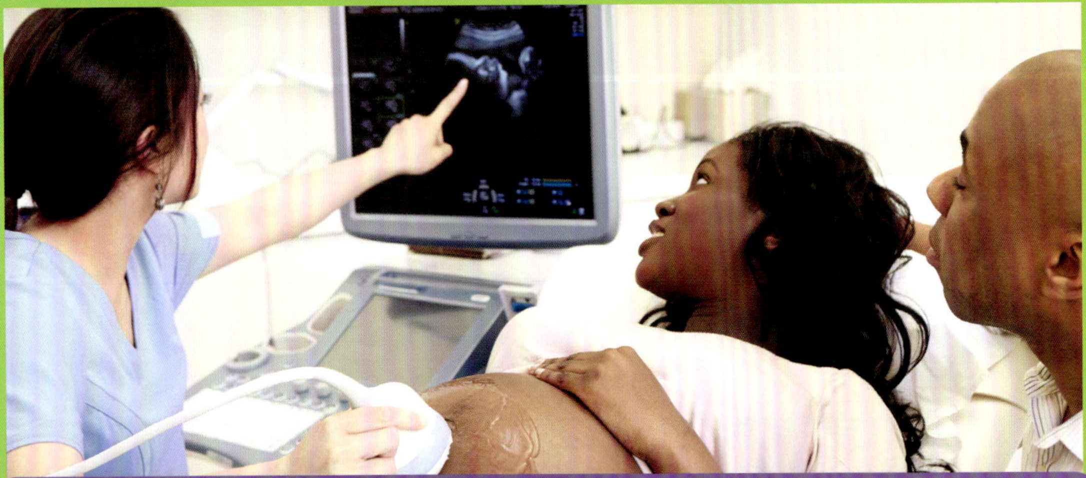

Ultrasound is a safe way for doctors to be able to examine a baby before birth.

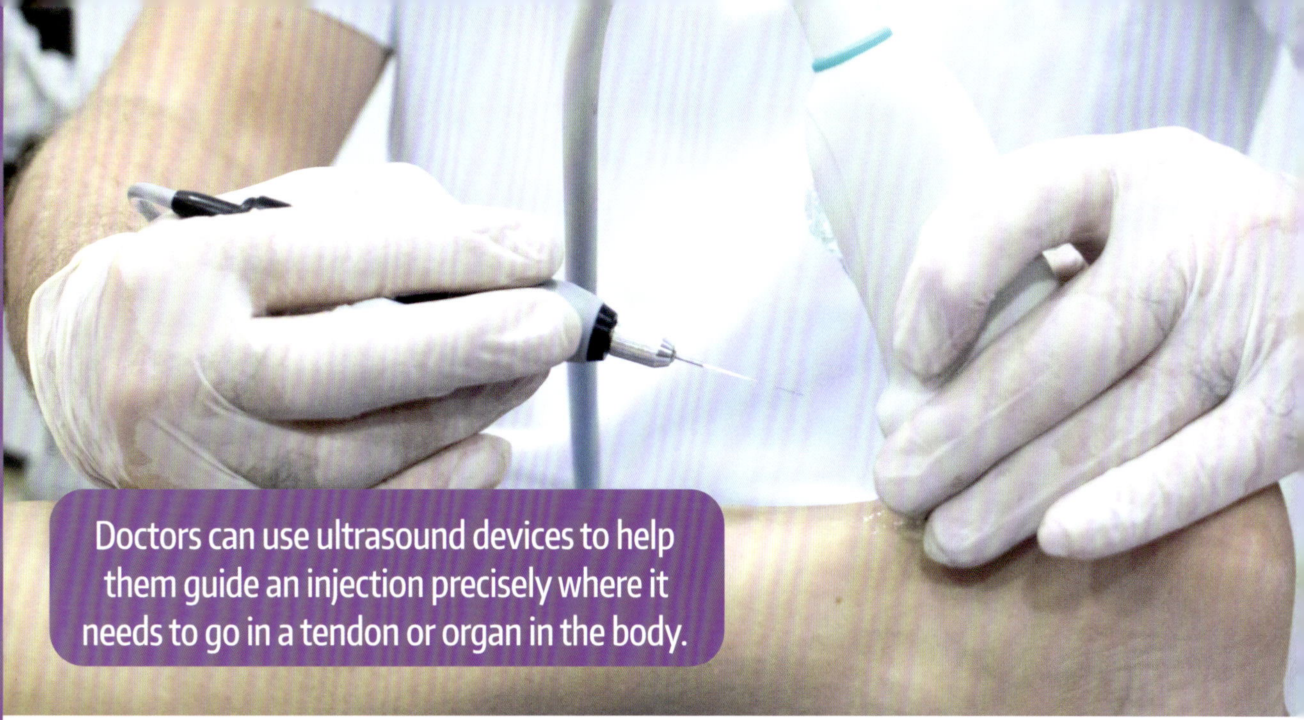

Doctors can use ultrasound devices to help them guide an injection precisely where it needs to go in a tendon or organ in the body.

Doctors are experimenting with ways to use sound to heal people, too. A new treatment uses high-frequency sound waves to kill certain types of cancer cells, with few side effects. Another treatment, called MIST, uses ultrasound waves pushed through salt water to help heal wounds that aren't getting better on their own. Some people even use ultrasound technology to improve their skin and look more youthful.

Consider This

Why is it safer for a doctor to use ultrasound rather than to guess what is wrong with a patient or to perform surgery?

Over the Distance

We haven't always been able to talk to anyone we want at any time of day. For most of history, people had to write letters to communicate with anyone far away. Beginning in 1876, people were able to talk to one another over great distances thanks to the telephone. The telephone can **transmit** sound over long distances.

When using a telephone, you speak into a microphone. Your voice is changed into an electrical signal and then sent to another phone. When the signal gets to the other phone, it is changed back into sound through a speaker.

Many people claimed to have invented the first telephone, but Alexander Graham Bell, pictured here in 1892, received the first U.S. patent for the device.

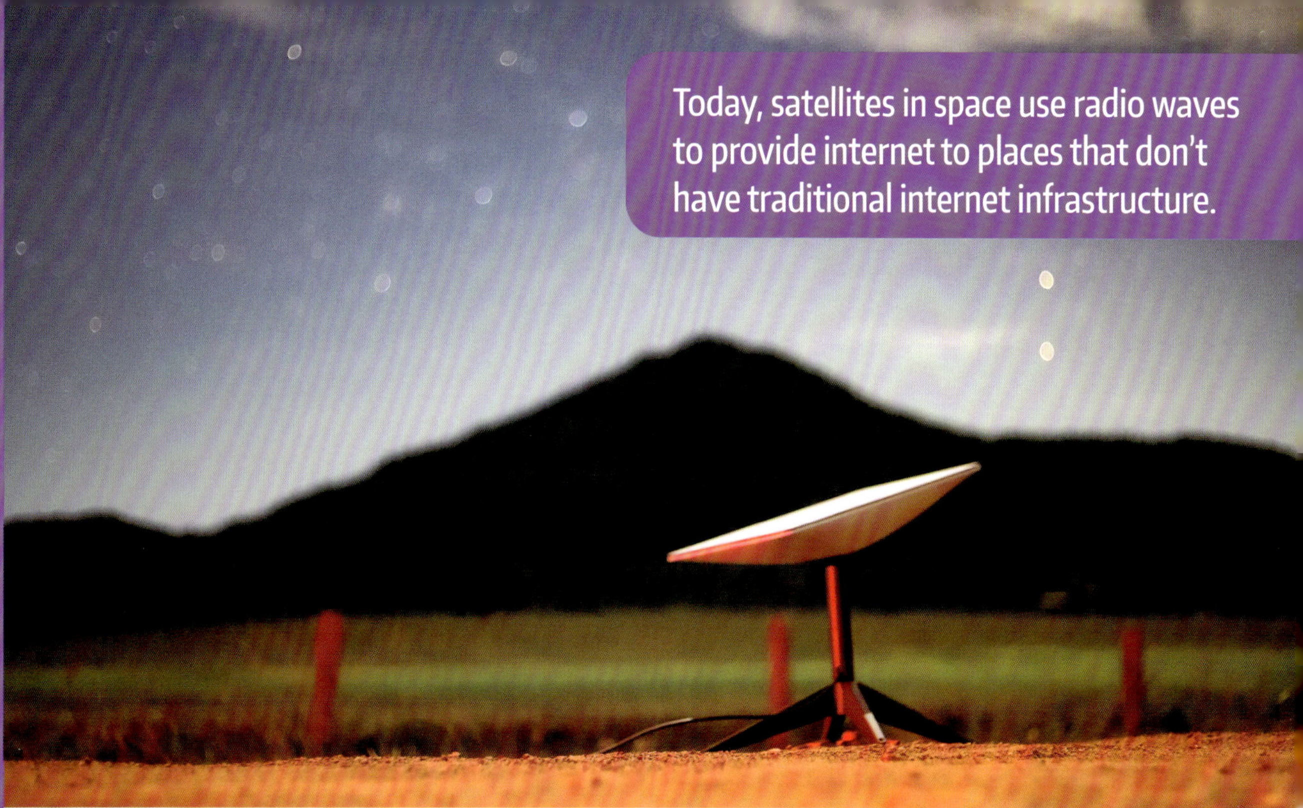

Today, satellites in space use radio waves to provide internet to places that don't have traditional internet infrastructure.

Telephones have changed over time. In early phones, the signals were sent over a series of wires and cables. Now, signals can travel through the air as radio waves. Cell phones, combined with receiving towers and satellites, let people speak with one another from any place in the world. Wi-Fi (wireless fidelity) uses radio waves to transmit information at high speeds. It is how many people access the internet, even in remote parts of the world.

WORD WISE

TO TRANSMIT IS TO SEND SOMETHING.

Keep Quiet!

Sometimes the goal isn't to amplify sound, but to reduce it. We can use certain materials to absorb or block sounds. Fabrics, such as carpets or drapes, can absorb some sounds. Insulation in walls can block noise as well.

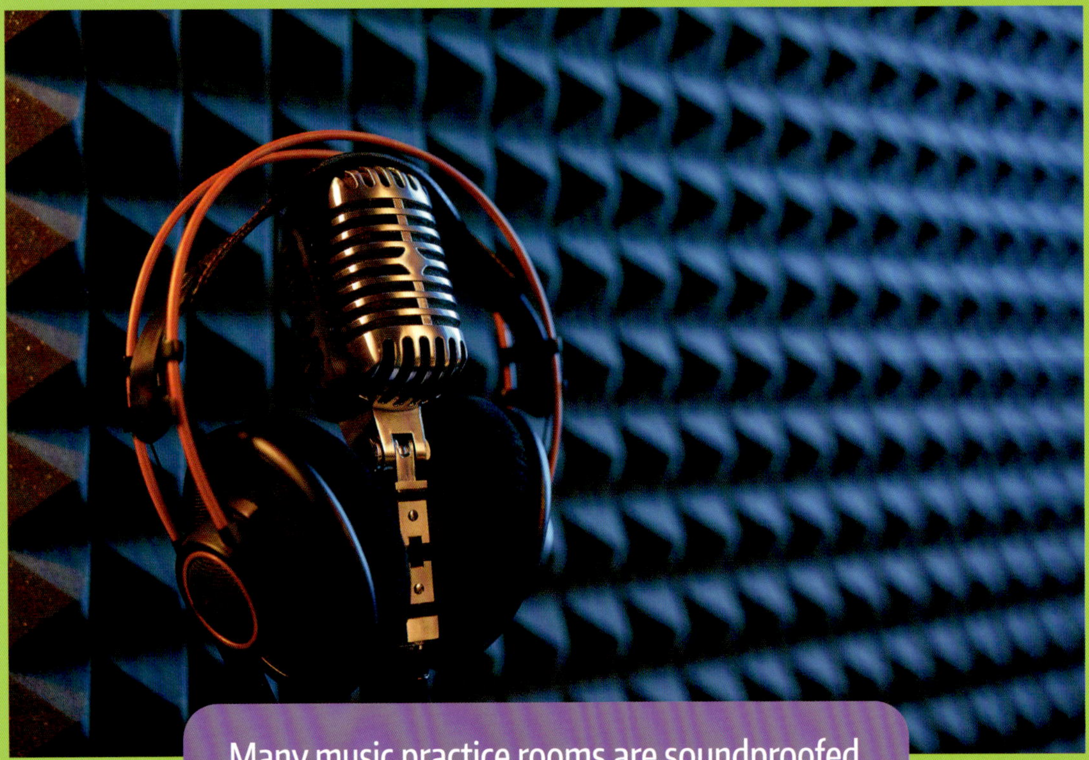

Many music practice rooms are soundproofed with foam or other materials that absorb sound.

People who work in loud environments, such as construction workers, often wear protection for their ears.

Noise pollution is a problem in big cities. Traffic and construction create lots of noise. Too much noise can make it difficult to sleep or concentrate and can even damage your hearing permanently. A sound at or above 120 dB is painful and can harm your ears. People who work with loud equipment usually wear earplugs to protect their hearing. When you use headphones, it's a good idea to keep the volume down. Once your hearing is impaired, it is very difficult to get it back. Consider wearing earplugs at loud concerts or even while watching fireworks shows.

Consider This

What are some other ways to reduce noise pollution and hearing loss?

Glossary

absorb To take in without giving back.

detect To discover or notice the presence of something.

functional Able to be used.

impaired Hurt or damaged.

infrastructure The equipment and structures needed for a country, state, or region to function properly.

laser beam A powerful beam of light.

manipulate To move or control something.

medium An object or material through which sound energy travels.

mimic To imitate.

navigate To find one's way.

particle A basic unit of matter that is very tiny and unable to be seen with the naked eye.

patent Rights given to a person by the government to make, use, or sell an invention.

receiver A device that takes in a signal.

satellite An spacecraft placed in orbit around Earth, a moon, or a planet to collect information or for communication.

side effects Things that happen to your body because of a certain treatment that are not meant to happen.

source Where something comes from.

vacuum A space that is completely empty of matter.

For More Information

Books

Gebhardt, Amanda. *Sound*. Ann Arbor, MI: Cherry Blossom Press, 2025.

Midthun, Joseph. *Sound*. Chicago, IL: World Book, 2022.

Williams, Victoria M. *Sound*. London, UK: Usborne Publishing, 2024.

Websites

Britannica Kids: Sound
www.kids.britannica.com/kids/article/sound/353791
Read more about sound energy and how it works all around us.

Ducksters: Basics of Sound
www.ducksters.com/science/sound101.php
Learn more about sound and explore the sound barrier and sonic booms.

Index